I0147346

Arthur Wingham

Report on the Analysis of Various Examples of Oriental

Metal-Work, &c.,

in the South Kensington Museum and other collections

Arthur Wingham

Report on the Analysis of Various Examples of Oriental Metal-Work, &c.,
in the South Kensington Museum and other collections

ISBN/EAN: 9783337218188

Printed in Europe, USA, Canada, Australia, Japan

Cover: Foto ©Thomas Meinert / pixelio.de

More available books at **www.hansebooks.com**

Department of Science and Art of the Committee of
Council on Education.

REPORT

ON THE

ANALYSIS

OF VARIOUS EXAMPLES OF

ORIENTAL METAL-WORK, &c.,

IN THE

SOUTH KENSINGTON MUSEUM

AND

OTHER COLLECTIONS.

MADE UNDER THE DIRECTION OF

PROFESSOR W. C. ROBERTS-AUSTEN, C.B., F.R.S.,

BY

ARTHUR WINGHAM, F.I.C.

LONDON:

PRINTED FOR HER MAJESTY'S STATIONERY OFFICE,
BY EYRE AND SPOTTISWOODE,
PRINTERS TO THE QUEEN'S MOST EXCELLENT MAJESTY.

1892.

Price Sixpence.

INTRODUCTION.

THIS report is the result of an investigation undertaken at the suggestion of Prof. Valentine Ball, C.B., F.R.S., with the view to ascertain whether any marked characteristic existed in the composition of Oriental brass from different localities, or from the same locality but of different ages, and it was hoped that it would be possible by the aid of an analysis to name the locality or to assign the time of production to a brass of uncertain origin, or at least to support any existing idea as to the date of a doubtful specimen.

It was necessary that the analyses should be as complete as possible, as a comparatively rough estimation of the principal metals only would be of little or no use. It was considered that a few exhaustive analyses would be more likely to yield definite results than a larger number of incomplete ones. Every specimen was therefore subjected to a rigid examination, and the greatest care was bestowed, at times under considerable disadvantages, on each analysis, so that every particle of metal could be accounted for and the quantity determined wherever there was sufficient present to admit of this being effected. The chief of the disadvantages referred to was the small quantity of the material at the analyst's disposal, which rendered the estimation of the metals present in small quantity a task of great difficulty, and one requiring special knowledge and skill.

In commencing the work, and keeping the main purpose well in view, it was considered useless to spend time in analysing any but authentic specimens. In the selection of these, the aid of Mr. C. Purdon Clarke, C.I.E., was sought and obtained in respect to the specimens from the Indian Section of the South Kensington Museum ; while the choice of those from the Persian Court was entrusted to Mr. A. B. Skinner.

Unfortunately very few brass or bronze objects in the Indian Section could be found whose authenticity as to date was beyond question, and these were nearly all comparatively speaking modern. A few brass specimens of doubtful age but of known locality were therefore chosen, as well as some objects of individual interest. Better fortune was met with in the Persian Court, where there are many authentic objects of various dates during the last ten centuries. A careful selection was made, with the result that the 15 analyses, Nos. 7 to 21 inclusive comprise a very interesting series of Persian objects, arranged in chronological order, extending over a period of from the 10th or

11th to the 19th century. These analyses show that the early Persian brasses and bronzes were indefinite mixtures, that the constituent metals were often impure, and further, that no definite composition was persistently aimed at at any one time, except perhaps in the specimens of the 19th century, which appear to be purer than the older ones, and which were probably worked up from imported English sheet brass. No regularity in composition could be traced. These results lead to the view that it would be impossible to judge with certainty the age of an alloy from its composition, which would, however, render it possible to distinguish between ancient and quite modern examples.

The question as to the possibility of determining the locality of specimens is not entirely disposed of. It would probably be difficult to name the province in which a brass was made, but a complete analysis might often help to decide between one of two districts. This would apply more particularly to the ancient specimens which would probably have been prepared from the ores found in the locality, before communication with other countries was as easy as it has been of late years, but a much more extended series of complete analyses, including all traces of metals, would have to be made before definite conclusions could be arrived at.

It is in connexion with this part of the subject that an instance is afforded of the value of carefully estimating the amounts of the metals present in small quantity. Thus it will be observed that in five out of the six Arabian specimens, Nos. 1 to 6, the percentage of lead is less than 0·7, while out of the 15 Persian specimens, Nos. 7 to 21, only three, Nos. 15, 18, and 21, contain less than 1·0 per cent. of lead, and of these only one (No. 15) contains no zinc, by the use of which in Persia probably would be introduced a considerable amount of lead. Another specimen (No. 21) is very likely made from imported English-made brass, so that it might be safe to conclude that if an old oriental brass contained less than 1·0 per cent. of lead, its brass was not made in Persia, and that in a question of choice between Arabia and Persia the probability would rest with it being Arabian. No. 18, (an Astrolabe) was, therefore, probably not of Persian, though it may possibly be of Arabian origin.

Besides the results to which reference has been made many objects to which especial interest was attached have been analysed, and the results in most instances fully repaid the work bestowed upon them. Among these may be mentioned the Thysius stoneware bottle (No. 58); "Hanuman," the monkey ally of Rama (No. 54); the specimens of Bells (Nos. 50, 51 and 52); the long Indian gun belonging to H.R.H. the Prince of Wales (No. 53); and the iron dowel (No. 57), which must be at least 700 years old.

There are also analyses of some specimens of Indian gold and a few objects of Indian Bidri work. The principal metals have been determined in some very ancient bronze objects in the British Museum, which were selected by Mr. A. S. Murray. The portions of metal analysed consisted of very small quantities of the filings which had been kept from the time the objects were being mounted on plinths for exhibition.

Attention may also be directed to an examination of Japanese art metal-work, the treatment of which was until recently enveloped in obscurity in European countries. The results show that the composition of the metal is less important in the production of the beautifully coloured patina so characteristic of Japanese work than was at one time supposed. The tint of the patina apparently depends mainly upon the composition and the manipulation of the various pickling solutions employed, a fact which by no means detracts from the difficulties of the art, or from the skill of the Japanese worker.

The descriptions at the head of the analyses are in most cases copies of the tickets exhibited with the objects in the Museum. It will be noticed that in a few instances the description of the metal does not agree with the results of the analysis.

These analyses do not support the opinion that the Eastern metal-workers prepared and employed exceptionally pure metals.

The following is the order in which the analyses, &c. have been arranged :—

Nos.

Specimens from the Arabian and Persian Courts of the South Kensington Museum—

Brass and bronze - - - -	1–21
Persian steel - - - -	22, 23
Specimens from the India collections—	
Gold and silver - - - -	24–35
Bidri - - - - -	36–39
Brass and bronze - - -	40–56
Iron dowel - - - -	57
Thysius bottle - - - -	58
Japanese metal-work from various collections	59–64
Japanese metal-work, coloured bronze plaques in South Kensington Museum - -	65–75
Experiments on Japanese patina, page 49.	
Pickling solutions used by the Japanese, page 51.	
Antique bronzes in the British Museum -	76–82

The following analyses have been made by Mr. Arthur Wingham who has admirably conducted the work entrusted to him.

W. C. ROBERTS-AUSTEN.

December 1891.

Analysis of Specimens taken from the Arabian and Persian Courts of the South Kensington Museum.

No. 1.

CANDLESTICK. For holding a wax candle in the tomb of a Saint. Bronze, nine-sided, incurved in the middle of the base, chased with inscriptions in Kufic letters, figures in medallions, and other ornament, partly inlaid with silver. *Persian.* 9th or 10th century. H. 10 in., diam. 9½ in. Bought, 5*l.* 10*s.*

571.-'78.

ANALYSIS.

Copper ..	..	..	..	79·77
Arsenic ..	..	..	..	Trace.
Antimony	..	..	..	Nil.
Tin ..	..	..	.	·141
Lead ..	..	..	..	·341
Bismuth..	..	..	..	Trace.
Iron ..	..	..	..	·100
Nickel ..	..	..	..	Trace.
Zinc, by diff.	..	..	..	19·648
Silver ..	..	..	..	Trace.
Gold ..	..	..	..	Nil.
				100·000

No. 2.

BOX AND COVER. Brass inlaid with silver, cylindrical, the edge of the cover bevelled and engraved with an Arabic inscription recording name and titles of El-'Adil Abú-Bekr II. (A.D. 1238–40), grandnephew of Saladin; the sides covered with aureoled figures, hunting scenes, &c., in chased silver inlay, enclosed in eightfoil dotted borders, on a ground of arabesques and flowers; on the cover, diaper of hexagrams

enclosing six seated turbaned figures round central sun, within a zone of the signs of the zodiac. On the bottom, an inscription records that it was made at the royal magazine. *Saracenic.* A.D. 1238–40. H. 4½ in., diam. 4¼ in. Bought, 60*l.*
8508.–'63.

ANALYSIS.

Copper ..	80·61
Arsenic..	·252
Antimony	Trace.
Tin ..	·126
Lead ..	1·441
Bismuth	Trace.
Iron ..	·252
Nickel ..	·182
Zinc, by diff. ..	17·070
Silver ..	·067
Gold ..	Nil.
	100·000

No. 3.

TRAY. Brass, circular; inlaid with gold and silver, ornamented with Arabic inscriptions on a ground of flowers, leaves, and birds. The inscriptions record the name and titles of the Memlúk Sultan En-Násir ibn Kalaún (A.D. 1293–1309) and are divided by panels of flowers and medallions containing either the Sultan's name, or an escutcheon, charge, an antelope within a fence. *Saracenic* (of Cairo). A.D. 1293–1309 Diam. 31 in. Bought, 40*l.* 420.–'54.

ANALYSIS.

Copper ..	81·20
Arsenic	Trace.
Antimony	·206
Tin ..	2·029
Lead ..	·457
Bismuth	Trace.
Iron ..	·147
Nickel ..	Trace.
Zinc, by diff. ..	15·886
Silver ..	·075
Gold ..	Nil.
	100·000

No. 4.

CANDLESTICK. Brass, inlaid with silver, and engraved with Arabic inscriptions recording Memlúk titles of the 14th century round base and neck and on other parts, divided by medallions of geometrical designs, and by rosettes of flowers. *Saracenic.* 14th century. H. 10 in., diam. 9 in. Bought, 6*l.* 8*s.* 4505.–'58.

ANALYSIS.

Copper	..	..	..	84·75
Arsenic	..	..	..	Trace.
Antimony	..	..	..	Nil.
Tin	..	..	..	2·360
Lead ..	..	..	..	·656
Bismuth	..	..	..	Trace.
Iron ..	..	..	..	·098
Nickel ..	..	..	..	Trace.
Zinc, by diff. ..	..	..	..	12·084
Silver ..	..	..	..	·052
Gold ..	..	..	..	Nil.

100·000

No. 5.

BOWL. Brass, inlaid with silver, and engraved with Arabic inscriptions, in a zone divided by four medallions, recording name and titles of the Memlúk Sultan Káït-Bey (A.D. 1468–96), on a ground of delicate arabesque ornament, and beaten out in a larger floral design engraved with scroll work. *Saracenic.* 15th century. H. 6¼ in., diam. 16½ in. Bought, 10*l.* 10*s.* 1325.–'56.

ANALYSIS.

Copper	..	..	..	75·11
Arsenic	..	..	..	Trace.
Antimony	..	..	..	Nil.
Tin ..	..	..	..	Trace.
Lead ..	..	..	..	·683
Bismuth	..	..	..	Nil.
Iron ..	..	..	..	·084
Nickel ..	..	..	..	·372
Zinc, by diff. ..	..	..	..	23·646
Silver ..	..	..	..	·105
Gold ..	..	..	..	Nil.

100·000

No. 6.

VASE, one of a pair. Brass, engraved with geometrical and arabesque ornament and Arabic religious inscriptions, and inlaid with silver dots. *Saracenic* (of Syria). 18th century. H. 12 in., diam. 6½ in. Bought, 12*l.* the pair. 379.–'80.

ANALYSIS.

Copper	..	..	..	67·15
Arsenic	..	..	..	Trace.
Antimony	..	..	..	Nil.
Tin ..	..	..	..	Nil.
Lead ..	..	..	..	·560
Bismuth	..	..	..	Trace.
Iron ..	..	..	..	·126
Nickel ..	..	..	..	Trace.
Zinc, by diff. ..	..	..	32·164
Silver ..	..	..	..	Trace.
Gold ..	..	..	..	Nil.

100·000

No. 7.

MORTAR. Bronze, octagonal, chased with arabesques and Kufic inscriptions. Round the sides are projecting knobs, and on one side is a ring suspended from a bull's head. Found in the ruins of the city of Rhages. *Persian.* 10th or 11th century. H. 5¼ in., diam. 7 in. Bought, 8*l.* 466.–'76.

ANALYSIS.

Copper ..	..	..	..	68·19
Arsenic	..	..	..	Trace.
Antimony	..	..	..	Nil.
Tin ..	..	..	..	3·327
Lead ..	..	..	..	14·990
Bismuth	..	..	..	Trace.
Iron ..	..	..	..	·448
Nickel ..	..	..	..	Trace.
Zinc, by diff. ..	..	..	12·963
Silver ..	..	..	..	·082
Gold ..	..	..	..	Nil.

100·000

No. 8.

MIRROR. Bronze, circular, the back ornamented in relief with two rampant sphinxes, addorsed, among floral scrolls; around these is a Kufic inscription. Various inscriptions and devices are engraved on the back. *Persian* or *Saracenic.* 11th (?) century. Diam. 4⅛ in. Bought, 1*l.* 10*s.*   442.–'87.

ANALYSIS.

Copper..	..	..	..	77·97
Arsenic	..	..	..	·094
Antimony	..	..	..	Nil.
Tin ..	..	..	..	5·215
Lead ..	..	..	..	6·866
Bismuth	..	..	..	Trace.
Iron ..	..	..	..	·084
Nickel ..	..	..	..	Trace.
Zinc, by diff.	..	..	..	9·019
Silver ..	..	..	..	·752
Gold ..	..	..	..	Trace.
				100·000

No. 9.

ASTROLABE ? Bronze, engraved. It appears to have been gilt, and is slightly inlaid with silver. It bears the date of the Hegira 598 (A.D. 1202), and was made at Damascus by Abdul Rahman, son of Iusuf. *Persian* or *Saracenic.* 9¾ in. by 8 in. Bought, 7*l.* 17*s.* 6*d.*   504.–'88.

ANALYSIS.

Copper..	..	..	..	79·40
Arsenic	..	..	..	·157
Antimony	..	..	..	Nil.
Tin ..	..	..	..	1·125
Lead ..	..	..	..	2·145
Bismuth	..	..	..	Trace.
Iron ..	..	..	..	·350
Nickel ..	..	..	..	Trace.
Zinc, by diff.	..	..	..	16·823
Silver ..	..	..	..	Trace.
Gold ..	..	..	..	Nil.
				100·000

No. 10.

BOWL. One of a pair. Brass, chased with inscriptions and diaper bands, partly inlaid with silver. *Persian.* 13th or 14th century. H. 4½ in., diam. 9¼ in. Bought, 4*l.* the pair.
559.–'76.

ANALYSIS.

Copper ..	..	..	..	75·74
Arsenic	..	..	..	·130
Antimony	..	..	..	Nil.
Tin ..	..	..	..	·306
Lead ..	..	..	..	2·268
Bismuth	..	..	..	Trace.
Iron ..	..	..	..	·196
Nickel ..	..	..	..	Trace.
Zinc, by diff.	..	..	..	21·360
Silver ..	..	..	..	Trace.
Gold ..	..	..	..	Nil.

100·000

No. 11.

LAMP STAND. Brass, cylindrical, chased with flowers, lozenges, and zigzag bands, and an inscription, filled in with black inlay. *Persian.* 13th or 14th century. H. 11¾ in., diam. 7½ in. Bought, 1*l.* 17*s.* 6*d.* 483.–'76.

ANALYSIS.

Copper ..	..	..	..	68·03
Arsenic	..	..	..	Trace.
Antimony	..	..	..	Nil.
Tin ..	..	..	..	5·986
Lead ..	..	..	..	5·950
Bismuth	..	..	..	Trace.
Iron ..	..	..	..	1·239
Nickel ..	..	..	..	·269
Zinc, by diff.	..	..	..	18·526
Silver ..	..	..	..	Nil.
Gold ..	..	..	..	—

100·000

No. 12.

WATER POT. Brass, with scroll handle and curved spout, inlaid with bands of inscription and arabesque ornament in silver. *Persian.* Dated, A.H. 866 (A.D. 1463). H. 6⅜ in., diam. 5⅛ in. Bought, 4*l*.   943.-'86.

ANALYSIS.

Copper ..	.. .. ..	75·32
Arsenic	.. .. ..	·063
Antimony	.. .. ..	Nil.
Tin ..	.. .. ..	1·022
Lead ..	.. .. ..	1·571
Bismuth	.. .. ..	Trace.
Iron ..	.. .. ..	·490
Nickel ..	.. .. ..	·144
Zinc, by diff. ..	.. ..	20·994
Silver ..	.. .. ..	·316
Gold ..	.. .. ..	·080
		100·000

No. 13.

POT WITH HANDLE. Bronze, chased with flowers and inscriptions. *Old Persian.* 15th century. H. 4⅜ in., diam. 7¾ in. Bought, 15*s*.   538.-'76.

ANALYSIS.

Copper ..	.. .. ..	73·19
Arsenic	.. .. ..	Trace.
Antimony	.. .. ..	Nil.
Tin ..	.. .. ..	6·317
Lead ..	.. .. ..	16·980
Bismuth	.. .. ..	Trace.
Iron ..	.. .. ..	·378
Nickel ..	.. .. ..	Trace.
Zinc, by diff. ..	.. ..	2·649
Silver ..	.. .. ..	·436
Gold ..	.. .. ..	·050
		100·000

No. 14.

MORTAR. Bronze, with cylindrical body encircled by raised bands, and a broad flange at top and bottom. The top flange is incised with an Arabic inscription and flowers, and the body with circular ornaments. *Persian.* 15th or 16th century. H. 7 in., diam. 10⅝ in. Bought, 2*l*. 15*s*. 948.–'86.

ANALYSIS.

Copper ..	.. .. ..	79·62
Arsenic	.. .. ..	·355
Antimony	.. .. ..	Trace.
Tin ..	.. .. ..	·519
Lead ..	.. .. ..	3·204
Bismuth	.. .. ..	Trace.
Iron ..	.. ?. ..	·112
Nickel ..	.. .. ..	Trace.
Zinc, by diff. ..	.. ..	16·190
Silver ..	.. .. ..	Trace.
Gold ..	.. .. ..	Trace.
		100·000

No. 15.

BOWL. Bronze, engraved with an inscription in Persian characters. *Persian.* 1511 A.D. Bought, 9*l*. 3*s*. 9*d*. 1191.–'54.

ANALYSIS.

Copper ..	.. .. ..	77·04
Arsenic	.. .. ..	Trace.
Antimony	.. .. ..	Nil.
Tin ..	.. .. ..	22·35
Lead ..	.. .. ..	·184
Bismuth	.. .. ..	Trace.
Iron ..	.. .. ..	·056
Nickel ..	.. .. ..	·091
Zinc ..	.. .. ..	Nil.
Silver ..	.. .. ..	·045
Gold ..	.. .. ..	Nil.
		99·766

No. 16.

BOWL. Light bronze, with incurved mouth, the outside engraved all over with figures in medallions, Kufic inscriptions, and diaper ornament. *Persian.* 16th century. H. 3½ in., diam. 7 in. Bought, 13s. 7.–'86.

ANALYSIS.

Copper..	..	..	..	75·80
Arsenic	..	..	..	Trace.
Antimony	..	.•	..	Nil.
Tin ..	..	..	..	1·211
Lead ..	..	..	..	2·514
Bismuth	..	..	..	Trace.
Iron ..	..	..	..	·210
Nickel ..	..	..	..	·068
Zinc, by diff. ..	..	..		20·145
Silver ..	..	..	..	·052
Gold ..	..	..	..	—

100·000

No. 17.

BOWL. Bronze, engraved outside with bands of inscription filled in with floral details. *Persian.* Apparently dated A.H. 1021 (A.D. 1612). H. 5¼ in., diam. 14½ in. Bought, 1*l*. 16s. 770.–'88.

ANALYSIS.

Copper ..	..	.. ..	..	70·51
Arsenic	..	..	..	·110
Antimony	..	..	..	Nil.
Tin ..	..	.. ..	..	·047
Lead ..	..	..	..	6·900
Bismuth	..	..	..	Trace.
Iron ..	..	..	..	·266
Nickel ..	..	..	..	·076
Zinc, by diff. ..	..	..	..	22·016
Silver ..	..	..	..	·075
Gold ..	..	..	..	—

100·000

No. 18.

ASTROLABE. Brass, round, chased, with five inner plates. It is dated 1074 A.H. (A.D. 1663). *Persian.* 17th century. Diam. 7¼ in. Bought, 6*l.* 15*s.* 530.-'76.

ANALYSIS.

Copper ..	. .	..	..	61·72
Arsenic	..	..	..	·079
Antimony	..	..	..	Nil.
Tin ..	..	..	..	Nil.
Lead ..	..	..	..	·409
Bismuth	..	..	..	Trace.
Iron ..	..	..	..	·308
Nickel ..	..	..	..	Trace.
Zinc, by diff.	..	..	..	37·484
Silver ..	..	..	..	Trace.
Gold ..	..	..	..	Nil.

100·000

No. 19.

ASTROLABE. Brass, circular, chased or engraved all over, with revolving perforated disc and five inner plates. Made by the famous Abdul Ahmeh, and dated A.H. 1127 (A.D. 1715). *Persian.* Entire L. 8 in., diam. 4¼ in. Bought, 12*l.* 2*s.* 5*d.* 458.-'88.

ANALYSIS.

Copper ..	..	..	..	73·96
Arsenic	..	..	..	·071
Antimony	..	..	..	Nil.
Tin ..	..	..	..	3·146
Lead ..	..	..	..	4·222
Bismuth	..	..	..	Trace.
Iron ..	..	..	..	·308
Nickel ..	..	..	..	Trace.
Zinc, by diff.	..	..	..	18·293
Silver ..	..	..	..	Trace.
Gold ..	..	..	..	Nil.

100·000

No. 20.

BASIN. "Lagan." Brass, with wide curved lip, chased with floral ornament. *Persian.* 18th century. H. 5 in., diam. 12½ in. Bought, 1*l.* 15*s.* 952 –'86.

ANALYSIS.

Copper..	..	..	..	75·15
Arsenic	..	..	..	·789
Antimony	..	..	..	Trace.
Tin ..	..	..	..	4·727
Lead ..	..	..	..	5·988
Bismuth	..	..	..	Trace.
Iron ..	..	..	..	·798
Nickel ..	..	..	..	Trace.
Zinc, by diff. ..	..	..	..	12·548
Silver ..	..	..	..	Trace.
Gold ..	..	..	..	Nil.

100·000

No. 21.

VASE WITH COVER AND HANDLE. "Dakhl-i-pul." Brass, pierced with floral designs, and chased with figures in medallions and a band of inscriptions. These vessels are suspended in shops in the bazaars, principally in the kebāb or roast meat shops, for the purpose of holding copper money. *Persian.* 19th century. H. 11¼ in., diam. 6½ in. Bought, 2*l.* the pair. 510*a.*–'76.

ANALYSIS.

Copper..	..	..	..	60·26
Arsenic	..	..	..	·086
Antimony	..	..	..	Nil.
Tin ..	..	..	..	Nil.
Lead ..	..	..	..	·915
Bismuth	..	..	..	Trace.
Iron ..	..	..	..	·112
Nickel	..	..	..	Trace.
Zinc, by diff. ..	..	..	..	38·627
Silver ..	..	..	..	Nil.
Gold ..	..	..	..	Nil.

100·000

• 66223. B

No. 22.

BOWL. Watered steel, with garlands, trellis ornament, &c. in gold inlay outside, and an inlaid inscription at the bottom inside. *Persian.* H. 3¾ in., diam. 8½ in. Bought (with 1304 to 1317), 36*l.*   1312.–'74.

ANALYSIS.

Carbon   ..   ..   ·94

No. 23.

BOTTLE WITH COVER, one of a pair. Steel, chased with flowers, birds, animals, and inscriptions, inlaid with gold. *Persian.* 19th century. H. 18⅜ in., diam. 5⅛ in. Bought, 24*l.* the pair.   383.–'80.

ANALYSIS.

Carbon   ..   ..   ·92

REMARKS ON ANALYSES.
(1 to 23.)

The foregoing results possess a collective value, and a close examination of them will bring to light many points of interest. The first six specimens are Arabian, then follow 15 Persian brasses and bronzes chronologically arranged, and finally two Persian steel objects. It will be noticed that the Arabian specimens, and especially the earlier ones, appear to be richer in copper than the Persian, and are of the nature of brass as we now understand it, *i.e.,* an alloy of copper and zinc. As previously pointed out in the introduction, five of the six Arabian specimens contain less than 0·7 per cent. of lead while only three of the 15 Persian analyses give low results. This would indicate that the Arabian zinc ores were purer, or, at any rate, freer from lead, than the Persian. Zinc as an isolated metal was not known in early times, and brass was made by melting copper with zinc ores or reducing copper ores in conjunction with those of zinc. Consequently any metallic impurities in the ores themselves would almost with certainty show themselves in the resulting brass. Lead as Galena is generally present in considerable quantities in the zinc ores known as Blende, which also often contain small amounts of iron and other metals.

There is also a marked difference in favour of the Arabian samples in respect to the per-centage of iron. In good modern brass, the iron is nearly always below 0·17 per cent. and the lead should be less than 0·5 per cent.; more than these quantities would indicate that a bad variety of zinc had been used in the preparation of the alloy. A considerable amount of iron might indicate one or both of two things :—either that impure zinc had been employed or that the brass had been frequently remelted and stirred or brought into contact with iron materials. The large quantity of iron, indicated by most of the analyses of the Persian specimens could not be accounted for by the latter explanation. Consequently its presence may with certainty be attributed to the fact that it existed as an impurity in the zinc ore from which the brass was made. This would be supported by the irregularity in the quantity present, and by the fact that in the bronze object (No. 15) where zinc is absent the per-centage of iron is very low.

A curious point about these analyses is the large quantity of silver occasionally present, especially in some of the older specimens, Nos. 8, 12, and 13. Gold is rarely present, and in only two cases, Nos. 12 and 13, is there a sufficient quantity to render its estimation possible. Nickel is generally present, but in small amounts, Nos. 5, 11, and 2 being the highest records. Arsenic as a rule is low, as it should be in brass, but it is very high in No. 20 and in Nos. 14 and 2, whilst there is over 0·1 per cent. in Nos. 9 and 10.

Antimony is generally absent, though sometimes a "trace" is present. It is a great enemy to brass, rendering it crystalline and brittle. The amount, therefore, in No. 3 is excessive, and the existence of only a trace of arsenic in conjunction with it is remarkable. This analysis is unique as showing a large predominance of antimony over arsenic, an unusual occurrence—the amounts being generally reversed. Another instance of this—a still more marked one—will be noticed amongst the Indian specimens (No. 40). Bismuth was only just detected in some cases by a very delicate test, and has been recorded as a trace in all the specimens (Nos. 1 to 21) except No. 5, where it was absent.

In respect to the principal metals of these alloys (copper, zinc, lead, and tin) there seems to be no regularity whatever in the relative amounts present. Of course the copper largely predominates, and is always over 60 per cent.; if less were present the metal would be rather unworkable. The zinc next predominates over the other two, except in one or two instances when the lead takes the second place; sometimes the tin predominates over the lead, and sometimes over the zinc, but it never ranks second to the copper except in the case of No. 15, which contains no zinc and very little lead, and is not a brass but a typical bronze, in which of course tin is the secondary metal.

The per-centage of copper is irregular, but on the whole high, omitting the very low examples. The amount generally present in modern brass for industrial purposes is between 66 and 75 per cent., the remainder being zinc with about 0·5 per cent. of lead which exists as an impurity in the zinc. "Yellow" or "Muntz metal" for ship sheathing, etc. contains 60 per cent. of copper. Ancient brasses possess a better colour than modern varieties, the majority of which would probably contain less than 70 per cent. of copper.

The superiority in colour of the older examples may be due entirely to the composition of the metal. In this report are a number of analyses which it would be of great interest to scrutinise in conjunction with the tints of the metals, and which might prove of much value to the modern art metal-worker in the selection of his metal.

The 19th century specimen (No. 21) looks very much like English yellow metal made with a bad or "dross" zinc—containing much lead and iron—an inferior brass not sufficiently good for home manufacturers and consequently exported to the East.

The last two specimens in this list, Nos. 22 and 23, are what are technically called high-carbon steels. They are as rich in carbon as die-steel and as some tool-steels. They are very hard, and a great amount of time must have been spent in working them.

Analysis of Specimens taken from the Indian Court of the South Kensington Museum.

No. 24.

BURMESE GOLD. Taken from various objects in the Royal Treasure from Mandalay, now in the India Museum.

ANALYSIS.

Gold	..	..	..	.. 90·70
Silver	..	..	..	.. 6·92
Copper	..	..	..	.. ·92
Iron	..	..	..	.. ·33
Platinum and other rare metals			..	1·12

99·99

No. 25.

THRONE, octagonal, on eight feet; gold plates chased and repoussé with bold floral and foliated ornament, mounted on wood; with three red and yellow velvet cushions. Made for Runjeet Singh after his accession to Lahore in 1799 A.D. H. 3 ft. 1¼ in., diam. 2 ft. 5 in.   2518.

ANALYSIS.

Gold	..	..	..	.. 97·75
Silver	..	..	..	.. 1·11
Copper	..	..	..	.. 1·01
Iron	..	..	..	.. Nil.

99·87

No. 26.

GOLD COIN. *Moghul.* 16th century. Transferred from the India Office.   Mus. No. ⅖.-1882.

ANALYSIS showed that it is of pure gold, as not a trace of any other metal could be detected.

Specially tested for silver, platinum, copper, iron, and lead.

Specific gravity of coin .. .. 19·265

No. 27.

GOLD COIN (*Indo-Scythian*).

Specific gravity -   -   - 18·698

ANALYSIS.

Gold	..	..	..	..	96·96
Silver	..	..	..	..	2·41
Copper	..	..	..	..	·64

100·01

No. 28.

RELIO SHRINE. " Stupa," in four parts (incomplete). Gold repoussé and chased with conventional ornament in bands. *Burmese.* Discovered in levelling a Buddhist Temple at Rangoon in April, 1855. H. 15 in., diam. 12¼ in.   02755.

ANALYSIS.

Gold	..	..	..	..	72·02
Silver	..	..	..	..	22·96
Copper	..	..	..	..	5·24

100·22

No. 29.

RELIC CASKET, cylindrical; gold, repoussé, set with two rows of rubies, and surrounded by an arcade containing figures of saints, the spandrils filled by birds with out-stretched wings; on the base a conventional lotus. Found in Afghanistan. *Buddhist.* 1st century B.C. H. 2⅝ in., diam. 2⅜ in. Lent by the Secretary of State for India in Council.

ANALYSIS.

Gold	..	..	..	..	90·85
Silver	..	..	..	..	5·63
Copper	..	..	..	..	3·65

100·13

No. 30.

RELIC SHRINE. "Stupa." Gold repoussé and chased with conventional ornament in bands, surmounted by a finial terminating in a bud ; set with 38 rubies and one emerald. *Burmese.* Discovered in levelling a Buddhist Temple at Rangoon in April, 1855. H. 14½ in., diam. 7¾ in.   02756.

ANALYSIS.

Gold	..	..	..	.. 94·72
Silver	..	..	..	.. 2·72
Copper	..	..	..	.. 2·75
				100·19

No. 31.

TASSEL. 16 stems, gold plates held by spirals of gold wire. *Burmese.* Discovered in levelling a Buddhist Temple at Rangoon in April 1855.   02753.

ANALYSIS.

Gold	..	..	..	.. 87·42
Silver	..	..	..	.. 8·44
Copper	..	..	..	.. 3·97
				99·83

No. 32.

RELIC SHRINE. "Stupa," in three parts (incomplete). Gold repoussé and chased with conventional ornament in bands. *Burmese.* Discovered in levelling a Buddhist Temple at Rangoon in April 1855. H. 2 in., diam. 3¼ in.   02754.

ANALYSIS.

Gold	..	..	..	.. 98·01
Silver	..	..	..	.. 1·57
Copper	..	..	..	.. ·38
				99·96

No. 33.

BOWL. Gold, unornamented, containing calcined bones and ashes. *Burmese.* Discovered in levelling a Buddhist Temple at Rangoon in April, 1855. H. 5 in., diam. 9¼ in.   02752.

ANALYSIS.

Gold	..	..	..	.. 83·07
Silver	..	..	..	.. 10·36
Copper	..	..	..	.. 6·43

99·86

No. 34.

HELMET, MODEL OF; probably a votive offering. Gold, with a double pin, and 62 jewels set in a border of repoussé ornament. Discovered in levelling a Buddhist Temple at Rangoon in April, 1855.   02758.

ANALYSIS.

Gold	..	..	..	.. 83·75
Silver	..	..	..	.. 12·22
Copper	..	..	..	.. 4·24

100·21

No. 35.

ANCIENT GREEK SILVER PATERA. Found by Dr. Lord in Badakshan. 4th century, A.D. Lent by the India Office.

ANALYSIS.

Silver	..	..	..	.. 93·63
Gold	..	..	..	.. 1·04
Copper	..	..	..	.. 4·94
Iron	..	..	..	.. ·27

99·88

No. 36.

EWER with COVER. Oxidised metal, damascened with silver.
Indian, modern. H. 2 ft. 11 in,, diam. 2 ft. 4 in.
Given by Her Majesty the Queen.   587.–'54.

ANALYSIS.

Lead	1 298
Copper	3·510
Iron	·049
Silver	Nil.
Zinc, by diff.	95·143
	100·000

No. 37.

BUDDHA, seated, praying; zinc. H. 4 in.   471.

ANALYSIS.

Copper	2·890	
Tin	·267	
Lead	·172	
Iron	·429	
Nickel	Nil.	
Silver	Nil.	
	3·758	Impurity.
Zinc, by diff.	96·242	
	100·000	

No. 38.

BOTTLE with COVER. Metal, with raised chasing of flower and
leaf pattern, covered with gold and silver foil. *Hungarian* or
Indian. Latter half of 17th century. H. 14⅜ in., diam. 5⅞ in.
Bought 10*l*.   356–'70.

ANALYSIS.

Lead	·956
Tin	·346
Iron	·084
Copper	Nil.
	1.386
Zinc, by diff.	98·614

No. 38A.

VASE. Oxidised metal, damascened with silver. *Indian,*
modern. H. 23 in., diam. 22 in. Given by Her Majesty the
Queen. 585 –'54.

ANALYSIS.

Lead..	..	..	..	..	1·437
Tin	..	:.	..	..	Trace.
Iron	..	..	..	..	·039
Copper	..	..	..	..	6·905
					8·381
Zinc, by diff.		..	..	..	91·619

No. 39.

WATER BOTTLE. Zinc, or a composite metal resembling it, of
elongated form, with flowers and animals in silver inlay, and
a foot and two bands of chased silver originally filled in
with cloisonné enamel. Resembling the Bidri work of India.
Persian? 17th or 18th century. H. 13¾ in., diam. 5 in.
Bought, 1*l.* 15*s.* 950.–'86.

From the Persian Court.

ANALYSIS.

Lead	..	..	..	..	·901
Tin	..	..	..	..	3·587
Iron	..	..	..	..	·476
Copper	..	..	..	..	·830
					5·794
Zinc, by diff.		..	..	..	94·206

No. 40.

VASE with HANDLE and COVER, " PHULDAN. " Perforated
brass; the cover surmounted by a knob; the handle formed

by two dragons. *Nepal.* H. 11½ in., diam. 7½ in. (Colonial and Indian Exhibition, 1886.) Bought, 2*l.* 10*s.*

89. (I.S.) 1886.

ANALYSIS.

Copper ..	..	..	..	78·58
Arsenic ..	..	..	..	Trace.
Antimony	..	..	..	2·037
Tin ..	..	..	..	·236
Lead ..	..	..	..	2·721
Bismuth ..	..	..	..	·070
Iron ..	..	..	..	·140
Nickel and Cobalt		..	..	·068
Zinc ..	..	..	..	15·849

99·701

No. 41.

LOTAH. Cast brass, with chased ornamentation. The body is ornamented with deeply indented intersecting rings. *Borneo.* Modern. H. 4 in., diam. 4½ in. Bought (Nos. 54 to 67), 10*l.*

54. (I.S.) 1886.

ANALYSIS.

Copper ..	..	..	..	59·80
Arsenic ..	..	..	..	Trace.
Antimony	..	..	..	Nil.
Tin ..	..	..	..	·298
Lead ..	..	..	..	3·226
Bismuth ..	..	..	..	Nil.
Iron ..	..	..	..	2·184
Nickel and Cobalt		..	..	·121
Zinc ..	..	..	..	34·53

100·159

No. 42.

PARESNATH, the twenty-third *Tirthankar,* or Saint of the *Jains;* impure brass. The figure has conventional features and hair, is nude except a girdle, and seated cross-legged on a triangular base, the front of which has conventional ornament in compartments, and is faced with copper and inlaid with silver. *Madras.* H. 2 ft. 2¾ in. W. 1 ft. 9 in. D. 1 ft. 1 in. 467.

ANALYSIS.

Copper	.. ..	69·96
Tin	.. ..	Nil.
Lead	.. ..	3·415
Bismuth..	.. .. ..	Trace.
Iron	.. ..	·504
Nickel ..	.. .. ..	Trace.
Zinc, by diff.	.. .. ..	26·121
Silver .•	.. .. ..	Trace.
		100·000

No. 43.

SHRINE FOR A JAIN SAINT, in three portions; brass with bronzed surface. A semi-circular arch supported by two subordinate ornamental shrines, each with a figure of an attendant bearing a chowrie. The bases and crown of the arch have each a small Jain temple enclosing the representation of a Saint. The arch bears a row of seated musicians, and within it is a star-shaped halo, and two elephants bearing musicians and worshippers. *Madras.* H. 3 ft. 8½ in. W. 3 ft. 2 in. 713.

ANALYSIS.

Copper ..	.. .. ..	75·83
Tin ..	.. .. ..	Trace.
Lead ..	.. .. ..	3·169
Bismuth ..	.. .. ..	Trace.
Iron ..	.. .. ..	·196
Nickel ..	.. ,. ..	Trace.
Zinc, by diff. ..	.. ..	20·805
Silver ..	.. ..	Cons. trace.
		100·000

No. 42 was considered to be copper or bronze, but on scraping was found to be brass. In scraping, a fine red dust came off which consisted largely of copper oxide, then the surface became more metallic with copper, on the removal of which the brass was exposed. The brass contained veins and blotches, apparently of copper.

No. 43. Brass, with bronzed surface, somewhat similar to above, but the metal was cleaner, softer, and better made.

These figures appear to have been originally all brass, from the surface of which the zinc has by some means been removed (by burning or by acid), and during which process the copper has become partly oxidised.

No. 44.

LOTAH, with RIBBED BOWL. Brass. *Calcutta.*   04,629.

ANALYSIS.

Copper	53·26
Arsenic	·205
Antimony	·126
Tin	2·312
Lead	3·983
Bismuth	Trace.
Iron	1·482
Nickel and Cobalt	·212
Zinc, by diff.	38·420
	100·000

No. 45.

GOGLET. Brass. *Travancore.*   04,644.

ANALYSIS.

Copper	76·70
Arsenic	·063
Antimony	Nil.
Tin	22·02
Lead	·839
Bismuth	Trace.
Iron	·196
Nickel and Cobalt	Nil.
Zinc	Nil.
	99·818

No. 46.

TANJORE BRASS STAND for offering Betel in Temples.
2269. (I.S.) 1883.

ANALYSIS.

Copper	71·39
Arsenic	·078
Antimony	Trace.
Tin	9·565
Lead	7·186
Bismuth	Trace.
Iron	·952
Nickel and Cobalt	·091
Zinc	10·459
	99·721

No. 47.

BULL, brass, on a base with incised ornament; a representation
of the *Nandi,* the sacred bull of *Shiva.*   *Benares.*   18th century.
H. 6 in., L. 5 in., W. 3 in.   2162.   (I.S.)   1883.

ANALYSIS.

Copper	66·09
Arsenic	Trace.
Antimony	Trace.
Tin	5·789
Lead	5·082
Bismuth	Trace.
Iron	·476
Nickel and Cobalt	·197
Zinc, by diff.	22·366
	100·000

No. 48.

BELL, brass, handle surmounted by a group of figures.   *Tanjore,*
   Madras.   02,930.

ANALYSIS.

Copper	77·14
Arsenic	·136
Antimony	Trace.
Tin	20·83
Lead	·410
Bismuth	·048
Iron	·093
Nickel	Trace.
Zinc	·398
Gold and Silver	Nil.
	99·055

No. 49.

BOWL, with figures and emblems in relief.   Brass.   *Java.*
                                                04,831.

ANALYSIS.

Copper	69·83
Arsenic	·197
Antimony	·031
Tin	9·192
Lead	19·08
Bismuth	·179
Iron	·252
Nickel	·180
Zinc	Nil.
	98·941

The above figures were all confirmed by repeated analysis. The sample was probably not clean metal and contained a quantity of oxide, which would account for the 1 per cent. discrepancy.

No. 50.

BELL, without clapper. *Burmah.*   9,291.

ANALYSIS.

Copper	..	..	..	69·55
Arsenic	..	..	..	Trace.
Antimony	..	..	..	Trace.
Tin ..	..	..	..	16·09
Lead ..	..	..	..	12·61
Bismuth	..	..	..	·054
Iron ..	..	..	..	·098
Nickel ..	..	..	..	·083
Zinc ..	..	..	..	1·556
Gold ..	..	..	..	Nil.
Silver ..	..	..	..	Trace.

100·041

INSCRIPTION (translation).

"In the month of Tabohdwe (February) on the fifth of the waning moon, in the year 1204 (1842, A.D.) on a Sunday at about 4 p.m. this bell was cast and moulded of pure copper. Its weight is 594,019 kyats (an obvious mistake). There are four lions on the hanging apparatus. Its height is nine fingers' breadths, the diameter five inches, the circumference fifteen, the thickness twenty-four. It is called the "Mahahtee Thadda Ganda" (the great sweet sound).

"The man who had this royal bell moulded was the Burman King, Tharrawahdy, Kohn Boung Min."

No. 51.

BELL, with dragons in relief and incised inscription. *Burmah.*
05,219.

ANALYSIS.

Copper ..	..	..	..	78·12
Arsenic	..	..	..	·221
Antimony	..	..	..	Nil.
Tin ..	..	..	..	19·18
Lead ..	..	..	..	1·913
Bismuth	..	..	..	·097
Iron ..	..	..	..	·105
Nickel ..	..	..	..	·114
Zinc ..	..	..	..	Nil.
Silver ..	..	..	..	·090
Gold ..	..	..	..	Nil.

99·840

The bell contained on its outer surface a long inscription, of which the following is part (translation):—

" We two, brother and sister, have given this bell as an offering to the seven Precious Things. The exact weight of the bell in current reckoning is 2,500 kyats. In this attempt to merit Neh'ban our arrangement was as follows:—We took our own weight in gold, and in silver and bright copper and other metal (Laukad, the Pali word used, implies five metals, gold, silver, copper, iron, lead), and mixed them well together. In the year 1209 (1847, A.D.) in the hot season, at a fortunate hour I had it moulded, setting my heart on giving it in alms.

" As I wrote the inscription I offered up abundant prayer, that no enemies or troubles might come nigh me, and that I might attain Neh' ban.

" Then I dedicated it."

No 52.

BELL, bronze. Swinging shackle ornamented with two grotesque lions. With inscription, dated 1828 A.D. *Burmese.* H 3 ft. 4 in., diam. 2 ft. 2 in. Presented by Commander C. McLaughlin, R.N. 76. (I.S.) 1884.

ANALYSES of Two SAMPLES, the first taken from the lip, the other from the top.

Copper ..	..	..	..	78·50
Arsenic	..	..	..	·205
Antimony	..	..	..	·193
Tin ..	..	..	..	12·73
Lead ..	..	..	..	5·520
Bismuth	..	..	..	·215
Iron ..	..	..	..	·322
Nickel }	..	..	..	Trace.
Zinc }	..	..	..	1·749
Silver ..	..	..	..	·263
Gold ..	..	..	..	Nil.
				99·702

Copper ..	..	..	..	78·05
Arsenic	..	..	..	·154
Antimony	..	..	..	·229
Tin ..	..	..	..	12·85
Lead ..	..	..	..	6·292
Bismuth	..	..	..	·251
Iron ..	..	..	..	·210
Nickel }	..	..	..	Trace.
Zinc }	..	..	..	1·524
Silver ..	..	..	..	·218
Gold ..	..	..	..	Nil.
				99·778

These two samples of the same object were taken to ascertain whether the metal was the same at one end of such a heavy and large Oriental casting as at the other, this being of especial interest as the amount of metal in the bell is very much more than could be melted in one pot, such as the Indian metal-workers could obtain. Weight, about 9 or 10 cwts.

It is well known that the Indians, when casting a large object, divide their metal into lots, which are placed and melted in small pots heated by separate fires, situated all round the foundry. When everything is ready each pot is taken out of its fire in turn and its contents poured into the mould. Consequently a large casting may have a very different composition at its opposite ends, especially if the alloy be a complex one. From the above analyses the metal in this instance appears to be more homogeneous than would have been expected.

A curious point about the analyses also is the tendency shown by the heavier metals* to predominate in the sample from the top, and *vice versâ* in that from the lip, distinctly pointing to separation. This would seem to suggest that the metal was cast in a hot mould and an attempt made to mix the various additions in the mould itself, with the consequent slow cooling, and, as a result, partial separation of the metals.

It would also tend to show that the bell was cast upside down, lip uppermost.

This bell contains a most interesting inscription full of pious sentiments, after which come the following extracts :—

In the city of Yathanahpoorah there lives a King, the Tshat-tan-seng · Meng - Shung - Bwah - Sheng- Meng-darah-gyu, who loves and honours God; . . . . and this King's devoted servant who can supply all the wants of his Sovereign instantaneously, who is gifted with all intelligence and bravery, and who also pursues the practices of a good man, first collecting all the property which he himself and his family have honestly acquired and then raising a subscription from all pious and good citizens, have with such means caused this bell . . . . of brass to be cast at the foot of the Tshoolay pagodah, and consecrated it to that pagodah.

For this good act in thus consecrating this bell, which sounds well, I pray that at a future period I may attain the same knowledge as God himself possesses, ... and thus become an Aruyah. I also pray that for this good act of mine the same blessing may attend my father, mother, teacher, the King, and the inhabitants of the three states. . . . . .

I call the Nuts (devils) of this earth to bear witness to this my act.

The Angel has placed his Seal.

I pray for knowledge of the past, of what exists, and of what is to come, or to see what is before me. Accomplished.

No. 53.

BURMESE GUN, in form of a dragon. Lent by H.R.H. Prince of Wales. Length 9 ft. 3 in., bore 3¾ in. 1887.

---

* Lead, bismuth, antimony, and tin.

ANALYSES of two samples, the first from the mouth end, the other from the tail or breech end.

Copper	..	..	..	.. 89·49
Arsenic	..	..	..	.. Trace.
Antimony	..	..	..	.. Trace.
Tin	..	..	..	.. 4·893
Lead	..	..	..	.. 2·118
Bismuth	..	..	..	.. Trace.
Iron	..	..	..	.. ·238
Nickel	..	..	..	.. Trace.
Zinc	..	..	..	.. 3·049
Gold ·	..	..	..	.. Nil.
Silver	..	..	..	.. Nil.

99·788

Copper	..	..	..	.. 85·93
Arsenic	..	..	..	.. Trace.
Antimony	..	..	..	.. Trace.
Tin	..	..	..	.. 7·654
Lead	..	..	..	.. 4·748
Bismuth	..	..	..	.. Trace.
Iron	..	..	..	.. ·210
Nickel	..	..	..	.. Trace.
Zinc	..	..	..	.. 1·501
Gold	..	..	..	.. Nil.
Silver	..	..	..	.. Nil.

100·043

No. 54.

HANUMAN; the monkey ally of *Rama;* copper and brass, cut through to show process of casting. *Madras.* Modern. H. 4 in.
726.

This specimen of art metal work is of peculiar interest, insomuch as it was apparently a compound casting showing on its surface two distinct metals, viz., bronze and brass, very much interspersed. The figure was not of one metal altered in colour in parts by superficial treatment, as inspection showed the colour to be due to the metal itself. Owing to the intermixture of the metals—the way in which they protruded one beyond the other at different points, and the small quantity at parts of one metal over the other such as the bracelets, armlets for example—it was difficult to understand from outside examination how these figures were made. With the object of solving the question it was considered that some light might be thrown on the subject if the casting were cut in halves from top to bottom. This was done, and immediately the whole process was explained, as it is quite clear that a core of copper was originally cast, of a shape showing due regard to the result desired, and that the brass was cast round the copper.

C

From what is known in connexion with small Oriental castings, it is probable that the following process has been employed in the production of these double castings. First, a model has been carved in wax, of a shape and size necessary to bring out the copper where that metal is required at the surface, and leaving space where the yellow surface is desired for the future casting of brass. This wax model has then been moulded, the mould heated and the wax melted out, after which the copper has been cast in the mould. Then round this half figure, with its prominent parts, where necessary, more wax has been cast, and carved into the shape of the figure ultimately required. The whole has then been moulded, the wax removed as before, and the brass run into the mould filling up the spaces existing between it and the copper core in the centre. The double casting has then been removed, the brass filed down, wherever it might accidentally and unnecessarily have covered the copper until the red metal was exposed, the whole being then chased and completed.

This casting is probably about 40 years old, and was made in Madras. The art of double casting, as represented by this figure, is very old (a thousand years or more), and has been practised at, and almost entirely confined to, the east coast of the Madras Presidency.

Mr. Havell, of the School of Art of Madras, who has been engaged on a survey of Art manufactures for the Government, reported in 1887, that these castings were no longer made. Mr. C. Purdon Clarke thinks that if this be true it is probably due to the high cost of production.

No. 55.

ONE of many coins found in the delta of the Nile. *Ancient Roman.* Lent by C. Purdon Clarke, Esq.

ANALYSIS.

Copper ..	..	..	..	81·84
Silver ..	..	..	..	17·74
Gold ..	..	..	..	·017
Arsenic..	..	..	..	·159
Antimony	..	..	..	·061
Tin ..	..	..	..	·042
Lead ..	..	..	..	·072
				99·931

No. 56.

TRAY OF KHORASSAN METAL. *Persian.* Lent by C. Purdon Clarke, Esq.

ANALYSIS.

Copper ..	..	..	..	50·37
Nickel ..	..	..	..	4·13
Cobalt ..	..	..	..	1·52
Iron ..	..	..	..	2·10
Lead ..	..	..	..	·38
Zinc, by diff.	..	..	..	41·50
				100·00

No. 57.

IRON DOWEL FOR STONEWORK. From the stonework of a temple destroyed by Altomash. · A.D. 1211–1236.

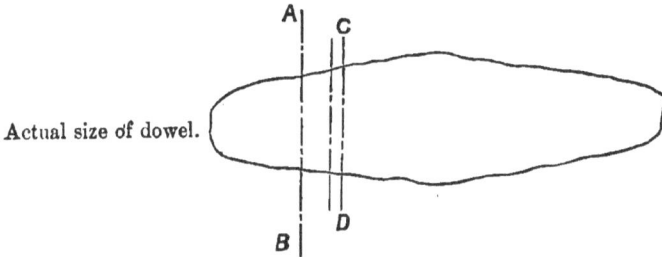

Actual size of dowel.

Section through AB was polished and etched with perchloride of mercury.

The etched surface showed the presence of a considerable amount of slag.

The section between the lines CD was taken for analysis.

ANALYSIS.

Silica ..	..	..	..	·20 = ·44 Slag.
Carbon ..	..	..	..	Trace.
Manganese	..	..	..	Nil.
Phosphorus	..	..	..	Nil.

The dowel is evidently part of a bloom hammered up into shape. The iron is practically pure, but the dowel contains about a half per cent. of slag, which had not been eliminated in working.

No. 58.

THYSIUS STONEWARE BOTTLE. *Egyptian.*

ANALYSIS.

Insoluble in hydrochloric acid.

Silica	..	..	..	65·32
Alumina..	..	..	..	19·22
Ferric Oxide	..	..	..	5·60
Lime	..	..	..	1·24
Magnesia	..	..	..	·68
				92·06

Soluble in hydrochloric acid.

Alumina..	..	..	..	1·24
Ferric Oxide	..	..	..	2·16
Lime	..	..	..	1·56
Magnesia	..	..	..	·30
Potash	..	..	..	3·06
				100·38

Amount insoluble in hydrochloric acid ..	92·64
Silica alumina, &c., as above	92·06
Amount of potash insoluble in hydrochloric acid	·58

The inside of the bottle contained an incrustation. This was examined. Mercury could not be found chemically or microscopically, showing that the bottle is not a mercury bottle.

The deposit consisted of chalk, with a small trace of nitrates and chlorine, which shows that this bottle has probably been used for carrying water. The deposit is not from sea water, as the chlorine is present in such small traces.

The bottle, from its composition, has evidently been made from a granitic alluvial deposit, and on examining microscopically crystals of black mica and felspar were seen.

HISTORY.

This bottle was one of a kind found in many parts of different countries, sometimes in large heaps, sometimes only two or three together. They are generally about four or five inches high, three or four inches broad, and have a very small orifice at one end.

They are somewhat of this shape, and are not made to stand.

The actual origin of these bottles has been very much discussed, and various uses have been assigned to them.

The bottle examined was found to be incrusted with a thin deposit of chalk, which indicates that this bottle was used for carrying water. The deposit was not, as might be supposed, the result of submergence under the sea and subsequent evaporation of the water, as the absence of anything but the merest trace of chlorine shows the chalk to have been derived from fresh water.

It has been suggested that these bottles could not be used as water bottles, as they have such a very small orifice, and they are supposed to have been used as mercury bottles. Both the incrustation and the bottle itself were examined most carefully for mercury, both chemically and microscopically, and not a trace could be found.

It has also been suggested that they might have been used for carrying the commercial articles of the day, such as saltpetre, or even used as bombs, but the absence of anything but a mere trace of nitrate renders both these suppositions improbable. The trace of nitrate, on the other hand, tends to support the view that they were used for water, as the quantity found was about equal in proportion to the chalk in the ratio often found in some fresh waters.

The bottle is very hard and cuts glass readily. A fracture showed a black or dark grey interior with yellow portions near both the inside and outside edges. The material had not been fused, but had been heated to a sufficiently high temperature to frit the mass together. From the regularity of the marks on the inside and outside of the bottle it is evident that some kind of potter's wheel has been used in turning the bottle into shape.

From the analysis the material used was apparently some alluvial deposit resulting from the decomposition of granite. Examining a powdered portion under the microscope, pieces of Felspar were plainly discernible, as also some rounded pieces of black mica and hornblende.

Finally, the suppositions arrived at are as follows :—

That the material from which the bottles were made was a river deposit from a granitic source ; in this instance, probably the bed of the Nile.

That the deposit was used as it was found, having been ground fine enough by the water's action.

That a kind of potter's wheel was used in turning the material into shape.

That the firing was sufficient to just frit the bottle material.

Finally, that the bottles were used for carrying water.

Prof. Judd has had a microscopical section made in his department, which, on examination, may throw further light on the geological question. He has also promised to obtain some Nile deposits, and it will be of interest to know whether or not they give the same peculiar dark grey appearance on being "fritted."

The following is the result of Prof. Judd's examination :—

"Felspar very distinctly present, and also a few crystals of hornblende. Plenty of quartz. All imbedded in a kind of glass."

"The material from which the bottle was made might possibly have been a deposit of the Nile. The section had a great resemblance to such deposits, and could very well be of that origin, but is not proof of such, although with other evidence it would tend to prove so."

"Nile deposits vary very much in their constituents, some being muddy, others less muddy with more sand."

A sample of Nile deposit, apparently muddy, was baked at a low temperature and gave a red cake due to excessive iron present. In a reducing atmosphere at a white heat the mass fused into a dark-coloured semi-glass. Thus it may be that with a deposit containing more sand a temperature might be obtained sufficient to fuse the mud, and thus, with the sand and other crystals of material present unfused but incased in the fused matrix, a material might be obtained very easily having the appearance of the bottle examined. From the fact that this bottle was found in Egypt also it is highly probable that the material from which it was made was a Nile deposit.

REMARKS ON ANALYSES.

(24 to 58.)

The analyses Nos. 24 to 58 need but little comment. The first 11 are specimens of gold, most of which appear to be native gold which little attempt has been made to purify. Native gold invariably contains silver, and very frequently copper. Some of the analyses show that means were adopted to purify the metal, especially No. 26, and Nos. 25, 30, and 32. All the rest are very impure.

No. 26 consisted of gold, the purity of which is very remarkable considering it is of the 16th century, an age when the methods of refining were somewhat crude. The specific gravity of pure gold varies according to its mechanical treatment between $19 \cdot 25$ and $19 \cdot 36$; this specimen was $19 \cdot 265$. The analysis No. 24 is an exceptional one as it indicates the presence of over $1 \cdot 0$ per cent. of platinum and other rare metals, and a small quantity of iron. It would be of interest to trace the source of this gold if it is possible to do so. The specimens Nos. 28 to 34 are all exhibited in the same glass case in the Museum, and the difference in colour between the rich and the poor specimens is readily discernible.

The Bidri objects, Nos. 36 to 39, are a peculiar series of alloys. They are chiefly zinc, containing a quantity of copper and lead with smaller amounts of tin and iron. The amount of lead in No. 37 is remarkably low, and, as it is very difficult to purify zinc from lead, it would appear that a very pure ore of zinc must have been used in the preparation of this metal. Apparently copper is the predominating metal associated with zinc in this Bidri alloy. From this it is probable that No. 38, which contains none of it, is not Indian but Hungarian, as the Museum description suggests. In the other specimen, to which a doubtful origin is assigned (No. 39), copper is present but in small proportion, tin being the secondary metal. This indicates that it is not Indian.

The remaining specimens are more of individual than collective interest. A few of the brass ones were selected as being derived from different localities, but their ages are uncertain, and they may or may not be typical alloys of the different districts. A greater number of specimens from each province would have to be examined before a definite decision could be given with certainty. The first thing noticeable in these analyses (Nos. 40 to 53) is the large quantity of bismuth present in some of them, and in one or two cases the extraordinary amount of antimony. The large amount of bismuth is in marked contrast to the Persian and Arabian analyses; with one exception, none of these alloys rich in bismuth are reported as coming from any district in the Madras Presidency, most of them coming from the north-east Provinces of India or Burmah. The alloys from these latter parts are altogether much more impure than those from the neighbourhood of Madras, indicating that the ores from the Himalaya districts must be rather complex.

The extraordinary quantity of antimony in the Nepal specimen (No. 40) renders it a matter of surprise that a metal with such a composition could be worked up into the shape of the object.

Iron is abundant in some of these metals, but it will be observed that it is generally low when the zinc is low or altogether absent ; so that its presence is probably due, as in the case of the Persian samples, to its presence as an impurity in the zinc ores used in making the brass. The per-centages of copper, zinc, lead, and tin, are very variable, and no deductions can be nor should be made from so few results.

The analyses of the Burmese Gun (No. 53), which is nine feet long, illustrate the difficulty of the Oriental native in obtaining a homogeneous alloy for a large casting. Samples were taken from each end of this long casting and analysed separately. The analyses differ widely one from the other, and show a want of uniformity in composition. Comparing them with those of other Burmese objects, it will be noticed that whereas Burmese alloys generally are very impure this metal is remarkably free from impurity. This clearly indicates that in producing this casting great pains were taken to purify the metals, or to select pure ores, so that a metal free from prejudicial impurities might be obtained. Neither of the alloys used is a particularly suitable one for a gun, and its want of uniformity would not improve the metal for the purpose for which it is required. Modern gun-metal generally consists of 90 per cent. of copper and 10 per cent. of tin.

Japanese Metal-work from various Collections.

No. 59.

CHOPSTICK CASE. *Modern Japanese.*

ANALYSIS.

Copper	..	..	..	.. 65·88
Arsenic	..	..	..	Cons. trace.
Lead	..	..	..	.. ·33
Bismuth	..	..	..	.. Trace.
Iron	..	..	..	.. ·37
Nickel	..	..	..	.. 6·14
Cobalt	..	..	..	.. ·50
Zinc, by diff.	..	..	..	.. 26·78
Gold	..	..	..	.. Trace.
Silver	..	..	..	.. Trace.
				100·00

This sample was an electro-plated one, and was covered with a coating of silver, which was removed prior to analysis. The surface scrapings contained much silver. The colour of the above metal was nearly white.

No. 60.

BRONZE TORTOISE. *Modern Japanese.*

ANALYSIS.

Copper	..	..	..	.. 81·62
Tin	..	..	..	.. 4·61
Lead	..	..	..	.. 10·21
Iron	..	..	..	.. ·22
Zinc	..	..	..	.. 2·42
				99·08

This sample was a very good and well-defined casting. Moderately hard and brittle.

No. 61.

SWORD-HILT. *Modern Japanese.*

ANALYSIS.

Copper	98·06
Arsenic	1·066
Antimony..	·010
Lead	Nil.
Bismuth	Nil.
Iron	·018
Nickel and Cobalt	·028
Zinc	Nil.
Tin	Nil.
	99·182

This sample was a chocolate brown colour, but was the ordinary copper colour on removal of the lacquered surface.

Nos. 62 and 63.

SWORD-HILTS. Iron, one round and one square, inlaid with Shaku-do. *Modern Japanese* Both from Prof. Roberts-Austen.

ANALYSIS, No. 62 (*round*).

Carbon	Trace.
Sulphur	·038
Silicon	·130
Phosphorus	Nil.
Manganese	Nil.

ANALYSIS, No. 63 (*square*).

Carbon	Trace.
Sulphur	·027
Silicon	·140
Phosphorus	Nil.
Manganese	Nil.

These analyses show that it is impossible that these sword-hilts could have been cast, as they are practically pure wrought iron with only traces of carbon, and therefore could not be melted at any temperature likely to have been used by the Japanese. This result was confirmed by the appearance of the etched polished surface of each, where the "lines of flow of the slag" proved that the metal had been hammered up.

No. 64.

MONKEY'S HEAD. Specimen of modern Japanese cast metal-
work. From Prof. Roberts-Austen.

ANALYSIS.

Lead	..	..	..	.. 92·39
Antimony..	..	..	..	7·61 by diff.

The above metal is coated with a very thin layer of copper, on the
surface of which the various effects are produced.

The brown colour is a patina of copper.

The eyes and teeth, which are of a golden colour, are not gold, but
Dutch metal or copper.

The back of the object is a white silver leaf, and consists of a thin
coating of silver deposited on the layer of copper.

Metal-work from Japanese Court, South Kensington Museum.

SPECIMENS of BRONZE. A collection of fifty-seven oblong plaques illustrating the different sorts of bronze, some of them with artificial patina, produced in Japan. Bought, 72*l*.
$$1099 \text{ to } \tfrac{1099}{57}.-'75.$$

Certain samples selected from the above were experimented on. The backs of the samples were carefully cleaned and scraped, and the scrapings were taken for analysis. The whole series were not analysed, but a few specimens were selected as being typical and characteristic, while two or three were chosen on account of a similarity in colour. These, upon analysis, gave results which showed that very varied colours could be produced from alloys, of which copper is the main constituent.

No. 65.

JIDAI-MEKKI-UTSUSHI. Imitation of old gilt bronze. Gold-coloured hammered surface plate. Museum No. $1099_{\frac{1}{23}}.-'75$. The back was copper colour, rather red. Metal moderately soft and free from cracks.

ANALYSIS.

Copper	99·63
Arsenic	·063
Antimony	Trace.
Tin	Nil.
Lead	·281
Bismuth	Trace.
Iron	·011
Nickel } Cobalt }	·021
Zinc	Nil.
Gold	Trace.
Silver	·042
Phosphorus	—
Sulphur	Nil.
	100·048

No. 66.

HIDO, Red Bronze. Bright, red-coloured, smooth surface plate. Museum No. 1099$_{\frac{2}{20}}$.-'75. The back was light copper colour. Metal rather hard and contained many cracks.

ANALYSIS.

Copper	..	..	..	97·87
Arsenic	..	..	..	Trace.
Antimony	..	..	..	Nil.
Tin..	..	..	..	·941
Lead	..	..	..	·328
Bismuth	..	..	..	Nil.
Iron	..	..	..	·077
Nickel ⎱ Cobalt ⎰	..	..	..	·043
Zinc	..	..	..	·546
Gold	..	..	..	Nil.
Silver	..	..	..	—
Phosphorus	..	..	..	Nil.
Sulphur	..	..	..	—

99·805

No. 67.

NIGURUMÉ. Black Bronze. Black-coloured smooth surface plate. Museum No. 1099$_{\frac{2}{26}}$.-'75. The back was ordinary copper colour. Metal moderately hard, and contained a few small cracks.

ANALYSIS.

Copper	..	..	..	99·25
Arsenic	..	..	..	·443
Antimony	..	..	..	·016
Tin..	..	..	..	Nil.
Lead	..	..	..	·197
Bismuth	..	..	..	Trace.
Iron	..	..	..	·011
Nickel ⎱ Cobalt ⎰	..	..	..	·016
Zinc	..	..	..	Nil.
Gold	..	..	..	Trace.
Silver	..	..	..	·039
Phosphorus	..	..	..	—
Sulphur	..	..	..	Nil.

99·972

No. 68.

NIGURUMÉ-NI-MATSUKAIVA. Light brown-coloured, coarsely hammered to resemble pine bark on a grey black bronze. Museum No. 1099$_{\frac{}{24}}$.-'75. The back was ordinary copper colour.

ANALYSIS.

Copper .. .. 99·37

No. 69.

NIGURUMÉ TSHUNÉ. Stone face on a black bronze. Dark brown-coloured, rough surface plate. Museum No. 1099$_{\frac{}{25}}$.-'75. The back was ordinary copper colour.

ANALYSIS.

Copper .. .. 99·26

No. 70.

KARAKANÉ. Ordinary Bronze. Dark brown-coloured, smooth surface plate. Museum No. 1099$_{\frac{}{27}}$.-'75. The back was apparently a light-coloured copper, but on scraping appeared quite white. Metal was very soft and easily cut. Scrapings were white, but tarnished to a light brown colour.

ANALYSIS.

Copper	73·27
Tin..	·865
Lead	20·344
Bismuth	Nil.
Zinc	5·460
Nickel	·060
Iron	·098
	100·097

No. 71.

SENTOKUDO. Yellow Bronze of Shinta. Brass, yellow-coloured smooth surface plate. Museum No. 1099$_{\frac{}{28}}$.-'75. The back was a light brass colour, somewhat similar to the front surface. Metal moderately soft. Scrapings yellow.

ANALYSIS.

Copper	..	..	..	.. 72·32
Tin ..	..	..	..	.. 8·126
Lead	..	..	..	.. 6·217
Bismuth	..	..	..	.. Trace.
Zinc	..	..	..	.. 13·102 (by diff.)
Nickel	..	..	..	.. ·065
Iron	..	..	..	.. ·170

100·000

From the above it will be seen that the appearance of the surface of the plates depends in a great measure on the subsequent treatment and not upon the composition of the metal. No. 65 is apparently gilt, as the existence of not more than a trace of gold in the metal prevents the possibility of the gold surface being due to the removal by solution of the other metals present. The same might be said to be the case with No. 67, where the black surface, which in some cases is known to be produced by chemical means, is here undoubtedly a surface coating only. No. 66 sample might have been coloured by oxidation, and afterwards lacquered, as the red colour is probably due to an oxidised compound of copper. As regards the composition of the metal in this sample the tin and zinc might have been added to harden the metal, as it is not likely that they were obtained simultaneously with the copper from the copper ore. The quantity of iron present suggests that the refined metal has been remelted at some time or other, and stirred with an iron rod probably when the tin and zinc were added. Or it may be that this metal was refined copper, to which some old bronze was added. The amount of lead present in each sample is probably due to the addition of that metal in refining the copper.

The amount of arsenic in No. 67 is large, and may possibly be due to the ore from which the copper was obtained, although the small per-centage of arsenic usually found in Japanese copper would not tend to support this.

The amounts of silver present are no more than are often found in copper from other sources.

As regards Nos. 68, 69, and 70, these were all of a brown colour, No. 68 being of a lighter shade, but Nos. 69 and 70 evidently the same colour; the other differences in these plates being the indentations or otherwise of the surfaces.

It may be that a particular alloy is used for No. 68, which is indented and coloured to represent the bark of a tree, but the determination of the per-centage of copper showed that the metal was only copper, containing the usual impurities. It was not deemed necessary to complete the analysis. The same with No. 69 sample. The sample No. 70, containing only 73 per cent. copper, was, however, completed, and showed an alloy of curious composition. It is probably refuse copper, from which the silver has been removed by the lead process. This sample was coloured the same tint of brown as No. 69, so that it is evident in this case that the composition of the metal has nothing to do with the colour of the surface. The Japanese apparently also do not specially select a soft metal for carving or indenting, as No. 70, which, from being soft, is well suited for indentation, is exhibited as a smooth surface, while the work of indentation is bestowed upon a harder metal.

No. 71 sample is of interest inasmuch as the colour of the lacquered surface is apparently due to the colour of the metal, and that the clean surface has simply been lacquered with a colourless lacquer. The above composition of the metal, however, is not necessary for the yellow colour, since it could be produced by copper and zinc alone. It is probably a chance alloy with a yellow colour, which yellow colour has been taken advantage of. This metal also is rather soft and suitable for easy indentation, but here again the advantage of the properties of the alloy in this respect has not been utilised.

The Japanese seemingly do not pay any attention to the composition of their metal in this work, and do not know with what they are working, except when they are using refined copper, and even then they are not particular if the surmise as regards No. 66 sample be correct. From the above few results it is evident that they produce their colours regardless of the impurities in the metal, which impurities it was at one time thought were the cause of the particular colour, and that they use and work with any metal that comes before them. When, during their re-meltings, they come across a metal which in itself is coloured they use it as such, as No. 71 shows, but in all other respects seemingly, for the purposes of colouring and lacquering the composition and properties of the metal are not in the least considered.

No. 72.

SHIBUICHI.   Greyish-white alloy. $1099_{\overline{\text{IT}}}$.–'75.

ANALYSIS.

Silver	..	..	..	..	37·91
Lead	..	..	..	..	·28
Iron	..	..	..	..	·16
Tin	..	..	..	..	Nil.
Nickel	..	..	..	..	Nil.
Zinc	..	..	..	..	Nil.

38·35

Analysis not completed.

Remainder   ..   ..   ..   61·65 probably all
                                         copper.

An analysis of a sword ornament of the finest quality, made by Mr. Gowland, late of the Imperial Japanese Mint at Osaka, gave the following results :—

Copper	..	..	..	..	67·31
Silver	..	..	..	..	32·07
Gold	..	..	..	..	Traces.
Iron	..	..	..	..	·52

99·90

Another analysis of this alloy, by Prof. Kalischer, proved it to contain :—

Copper	..	..	..	..	51·10
Silver	..	..	..	..	48·93
Gold	..	..	..	..	·12

100·15

No. 73.

KODO. Yellow bronze. 1099$\frac{1}{16}$.–'75.

ANALYSIS.

Copper	68·99
Arsenic.. ..	Small quantity, not determined.
Antimony	Nil.
Tin	Trace.
Lead	·888
Bismuth	Trace.
Iron	·196
Nickel	·098
Zinc, by diff.	29·828
Silver	Nil.
Gold	Nil.
	100·000

No. 74.

KODO-SOKOTSUCHIMÉ. Lateral hammer-work on yellow bronze.
1099$\frac{1}{16}$.–'75.

ANALYSIS.

Copper	69·72
Arsenic.. ..	Small quantity, not determined.
Antimony	Nil.
Tin	Nil.
Lead	·915
Bismuth	Trace.
Iron	·154
Nickel	Trace.
Zinc, by diff.	29·211
Silver	Nil.
Gold	Nil.
	100·000

No. 75.

KIODO. Sonorous bronze. 1099$\frac{1}{14}$.–'75.

ANALYSIS.

Copper	74·19
Arsenic..	Trace.
Antimony	Nil.
Tin	9·732
Lead	5·492
Bismuth	Trace.
Iron	·434
Nickel	Trace.
Zinc, by diff.	10·152
Silver	Nil.
Gold	Nil.
	100·000

Experiments on Japanese Patina.

ANALYSES OF COPPERS containing small quantities of impurities used in pickling experiments:—

Copper and gold	-	·465 Au.
Copper and silver	-	·402 Ag.
Copper and arsenic	-	·248 As.
Copper and antimony	-	·254 Sb.
Copper and lead	-	·355 Pb.
Copper and bismuth	-	·475 Bi.
Copper and iron	-	·213 Fe.

Pure copper used for the above contained 99·65 Cu. compared to electrotype copper of tested purity used as a standard. Difference probably due to dissolved oxide.

Polished samples of these alloys when heated in the solutions (Nos. 1 and 11) given on p. 53, showed that a remarkable variation of tint is produced by the presence of impurity, but it would require coloured illustrations to demonstrate this.

GOLD PATINA,

Containing shining pieces of metallic copper, and cakes of a greyish purple dried powder, evidently the true patina.

Weight taken - - -	·0442 gram.	
Wt. after two hours at 100° C. -	·0442 „	
„ „ ignition - - -	·0490 „	
„ „ dissolved in dil. HNO_3,⎫ apparently gold -⎭	·0033 „	
„ gold after sol. and pption. -	·0027 „	
„ copper oxide after pption. -	·0453 „	
„ residue insol. in aqua-regia -	·0006 „	

ANALYSIS.

Gold	..	..	..	6·11
Copper	..	..	..	81·9
Insol. res.	..	..	..	1·3
Chlorine	..	..	..	·4 (?)
Sulphur	..	..	..	Nil.

EXPERIMENT to extract any metallic gold by means of mercury.

·0389 grm. of patina was rubbed and pressed through about 1 grm. of mercury, then heated under water to about 60° C. for some time, excess patina washed away and mercury filtered thoroughly, dissolved in acid and solution tested for gold.

Not a trace had been dissolved.

The Japanese have a remarkable series of alloys, of which Shakudo is the most important, in which the precious metal gold and silver replace the tin and zinc of ordinary bronze. Analysis of a sword ornament of the finest quality by Mr. Gowland gave :—

Copper ..	..	..	..	94·50
Silver ..	..	..	..	1·55
Gold ..	..	..	..	3·73
Lead ..	..	..	..	·11
Iron and arsenic		..	..	Traces.

99·89

The amount of gold is, however, sometimes as high as 4 per cent., whilst silver is frequently only present in very small quantity. Shakudo when boiled in suitable pickles, such as No. III. of those given on p. 54, assumes a beautiful purple tint, and the following analyses have been made with a view to ascertain the composition of the purple patina.

EXPERIMENT with Chlorine Water.

·49 (?)	gold extracted by cold dilute Cl. water.				
·24 (?)	,,	,,	,, hot	,,	,,
4·92	,,	,,	,, hot conc.	,,	,,

*5·65 ,, ,,

This sample of patina is very much richer in gold than the metal from which it was made, and the patina itself is no doubt richer still than shown in above analysis, as the sample there indicated contained scrapings of metallic copper.

The figures marked with the query (?) are only approximate on account of the extremely small quantity weighed in each case.

The patina may possibly be a sub-oxychloride of copper and gold, or more probably a sub-oxide of copper and gold containing a small proportion of chlorine as sub-oxychloride.

---

* From these figures the following deductions are made :—
  That there may be some free gold present in the patina.
  That it is not proved that free gold is present.
  That if it be present the quantity is not more than ·5 or at the outside ·7 per cent.
  That there is a large quantity of gold present in the combined state.

GOLD PATINA, 2ND SAMPLE.

This sample, unlike the previous, consisted entirely of patina, as it had peeled off the copper on which it was formed. No separate pieces of copper could be seen. Under the microscope the outer surface of the patina had a grey metallic lustre, with a tinge of purple, but the inner surface was copper-coloured.

ANALYSIS.

Gold	..	..	..	8·54
Copper ..	..	..	...	80·21
Chlorine	..	..	..	·42
				89·17
Remainder	..	..	..	10·83 oxygen.

No free gold could be detected.

It will be seen that in this instance the gold is higher and the copper lower than before, while the chlorine (this time estimated on a larger quantity) confirms the previous result.

Working out a formula from the above figures, it is remarkable how closely it comes to a mixture of $Cu_2 O$ and $Au_2 O$.

Without concluding that the patina is always of the same composition, or that there is a definite formula for it, these latter results seem to support the previously expressed opinion that the patina is a sub-oxide of copper and gold containing a small proportion of chlorine as sub-oxychloride.

PICKLING SOLUTIONS USED BY THE JAPANESE.

—	I.	II.	III.
Verdigris - -	438 grains	87 grains	220 grains.
Sulphate of copper -	292 ,,	437 ,,	540 ,,
Nitre - - -	—	87 ,,	—
Common salt - -	—	146 ,,	—
Sulphur - -	—	233 ,,	—
Water - -	1 gallon	—	1 gallon.
Vinegar - -	—	1 gallon	5 fluid drachms.

The above are taken from a paper printed in the Journal of the Society of Arts, June 13, 1890.

The solutions are generally used boiling.

Antique Bronzes in the British Museum.

No. 76.

BRONZE FIGURE, with hand extended, as in Morra. Middle of
5th century B.C. Handle of a mirror. '73. 8–20. 19.

ANALYSIS.

Copper	89·96
Tin	7·64
Lead	1·44
	99·04

Zinc, Iron, Silver—traces.

No. 77.

BRONZE FIGURE OF DIONYSOS. Græco-Roman period.
'73. 8–20. 26.

ANALYSIS.

Copper	85·05
Tin	10·35
Lead	4·62
	100·02

Iron—trace.

No. 78.

BRONZE FIGURE OF A GLADIATOR. 3rd century A.D.
'73. 8–20. 53.

ANALYSIS.

Copper	79·26
Tin	4·71
Lead	7·05
Zinc	6·80
	97·82

Iron—trace.

No. 79.

BRONZE FIGURE OF A LION.   Etruscan work.   5th century B.C.
'73. 8-20. 251.

ANALYSIS.

Copper	82·10
Tin	12·64
Lead	1·86
Zinc	·73
	97·33

Iron—trace.

No. 80.

BRONZE FIGURE OF APOLLO, from Orange.   1st century A.D. (?)
'77. 8-10. 1.

ANALYSIS.

Copper	80·70
Tin	6·44
Lead	9·97
	97·11

Zinc, Iron, and Silver—traces.

No. 81.

BRONZE GROUP OF M. AURELIUS AND FAUSTINA.   2nd century
A.D.   '78. 3-9. 1.

ANALYSIS.

Copper	70·41
Lead	2·44
Zinc	26·70
	99·55

Tin, Iron—traces.

No. 82.

FRAGMENT OF DRAPERY found with leg of bronze statue. Greek sculpture. About B.C. 450. '86. 3-24. 7*a*.

ANALYSIS.

Copper ..	..	..	..	84·49
Tin	..	..	..	9·47
Lead	..	..	..	5·31
				99·27

Iron—trace.

These seven analyses—76 to 82—were all carried out on extremely small quantities of material, only 4 or 5 grains being received in some cases as the sample. Moreover, the samples of drillings were very dirty and partly oxidised. Hence the low totals obtained. The samples, however, were as carefully and thoroughly cleaned as possible before analysis.

The estimation of the metals present in small quantity was out of the question.

www.ingramcontent.com/pod-product-compliance
Lightning Source LLC
Chambersburg PA
CBHW031802090426
42739CB00008B/1127